Always Something More

Index of essays

Preface	2
Bridging The Gap	6
1. Osmosis	12
2. The Big Bang	15
3. Order And Similarity	22
4. Infinite Diversity	28
5. An Evolutionary Core	32
6. The Atom	37
7. Electromagnetic	43
8. Homo Sapiens	47
9. Gravity	55
10. Procreation	59
11. Dying To Live	66
12. Energy	75
A Conversation With ASM … Let's Share	80

Preface

I am ASM. My name is unfamiliar to you since it does not exist as an individual name or word in the English language. I intend ASM as a *spiritual/personal* name since it refers to something that ultimately defies measurement and description. I have chosen this name in order to capture the unfolding, dynamic awareness that: there is <u>A</u>lways <u>S</u>omething <u>M</u>ore --- **ASM**.

The initial vowel "A" in my name is pronounced like the "a" in "awesome!" Look at this word and say it out loud so that you can fully experience the wonder behind the word itself. Let it's breathy "otherness" infiltrate your entire being.
- Like entering a nice warm shower after a tense and uptight day and breathing out "ahhhhhh".
- Like standing in the presence of majestic, snow-caped mountains and breathing out "ahhhhh"
- Now breathe out my name adding the letters S and M. ASM = (awesome)

Young people frequently say that something is *Awesome*. Wikipedia describes this word for *something extremely breathtaking, awe-inspiring, magnificent, wonderful, amazing, stunning, staggering, imposing, stirring, impressive"*. Thus, I have chosen to capture the *essence* of this awareness in a single word that could become my name. I am deliberately choosing to add a personal dimension to it for I am ASM, the essential core at the center of all that exists

A name usually seeks to define something about a thing or person's character or nature. There are over 16 different names for "God" in Scripture. Each name attempts to reveal something about the essence of what you mean when you say "God" and you have inevitably tended to form a mental image that would correspond in some way when you use that word.

Some of the images that pop into your mind when you hear the word "God" are:
- *God is an elderly male,* frequently with a long white beard, infinitely wise and overseeing all things.
- *God is the omnipotent protector* who will keep you from all harm, and you have a right to be angry when God fails to do so.
- *God is a scorekeeper* in the heavens above who is keeping record of all you say and do, and preparing either reward or punishment in terms of heaven or hell.
- *God is a resourceful provider* capable of being enlisted (manipulated?) to fulfill your whims and desires if you pray in the right way.
- God is a *judge* who will mediate a decision of innocent or guilty on the final day.
- *God is Someone, Something, Somewhere, too great to comprehend, so why even try.*

I could go on and on with the varied images residing in your minds and hearts that give hints about something that you assume is real and part of the essence of God.

Consequently, the image you carry with you affects your awareness of what life is really all about and how you should subsequently act. In the past you have used the term "God", as viewed from your personal religious and experiential situation. All too often you have implied that your understanding of the word "God" is the only valid and proper understanding. Most of your human attempts to define the "God" you believe in is so loaded with baggage from your past that you are seriously distorting the essence of who I essentially am; so I am choosing a new name, not to replace the old ones, but to add to your list in order to stretch your understanding of who I AM.

I am ASM, THE REALITY that is not modified by anything, but rather, present in everything that exists and endlessly expanding into the wonder-filled unknown of what is yet to be.

Every time you find yourself filled with awe and sense the presence of an energy force that entices you ever onward, speak my name, ASM, loudly or softly and you will experience more and more of the essence behind my name.

Stay alert! Everything I am attempting to describe and say is *a metaphor* … a literary **figure of speech that uses something tangible to represent something less tangible but more comprehensive.** The essays I am sharing with you will inevitably go beyond your current awareness! Each essay contains paradox, mystery, and endless wonder, representing something crucial that is ultimately real but cannot be totally grasped and comprehended. To read each essay is to participate in a journey, knowing in advance that you cannot arrive at a destination! I welcome you to join me by experiencing an awareness that goes beyond what is normally received and assumed and contains the wonder-filled <u>A</u>lways <u>S</u>omething <u>M</u>ore that is my essence.

BEWARE! A word of caution. Nothing that I am going to share with you will become evident to you and speak to your inner being unless you enter into what I define as your *ASMize:* which is to relax, breathe deeply and then rhythmically, choose to be open-minded and reflect and meditate, looking for insights and nudges that surface within you. Then you must respond by actively applying the insight and nudge. You learn more about *the ASMize* under my essay of the same name. *Only through application and experience can you increasingly discover what is wonderful and true.* This exciting journey never grows boring and is always expanding because there is Always Something More. You will discover my "personality" as we proceed on this journey together.

Truth is always self-authenticating!

I repeat, insight and truth are always self-authenticating through the process of application! By taking whatever is spoken or written and applying it in your life you discover through the process of application if it is resonant and true!

It is my hope that you will find herein some "AHA" moments, which, when applied, will resonate with the essence of who I truly am and the essence of who you are truly meant to be. So, let the journey begin. ASM

Always Something More

Introduction -- Bridging The Gap

Science or Religion , Material or Spiritual,
Secular or Sacred,

Unfortunately, the period of history and culture in which you live tends to assume a vast chasm between:
- *science or religion.* *Science,* can be measured, analyzed, tested, and proven in a repeatable way and *religion* which deals largely with philosophical and moral concerns that defy being repeatable and measurable.
- *natural, or spiritual. Nature*, deals with the material world around us that can be analyzed under a microscope or through a telescope, and *spiritual* defies detailed measurement and explicit conclusions.
- *secular or sacred. Secular* is the surface perception of what is real and applicable and practically applied in this world; and *sacred* which is something which we call holy and assume to be vastly different and otherworldly.

Part of the chasm is that you have erroneously assumed that being secular means to be meaningfully involved in the world and anything sacred is otherworldly and impractical. You have erroneously assumed that an intellectual and scientific approach is far superior to religious concerns. And you have failed to recognize the inherent unity between that which is natural and that which is spiritual. That is why I have chosen to deal with this dichotomy in these lectures.

An important aspect of the tension that seems to exist lies in your attempt to interpret a Holy Book, "Scripture", only from a scientific and rational point of view and ignore the fact that your full awareness of reality reaches far beyond the intellectual, analytical mind alone that is communicated through *sight, sound, taste, touch and smell.* When mind is isolated and limited to a rational thought process, you can never fully comprehend the full reality of what is *essentially* present.

The Holy Bible is a compiled library of 66 books composed over a period of 2000 years by an unknown number of authors. The original copies are not in existence and their original composition was oral in nature and the exact insights and feelings behind the words used at the time they were spoken relates directly to the cultural circumstances historically present at that time. This is true of almost every religion and respective sacred book. The books are composed mostly of stories and historical occurrences and experiences that are *metaphorical, s*eeking to define a reality that reaches into the core of ones being and claims heart, soul, mind, and strength as a total reality and the essence of who I am, ASM -- the Always Something More.

Through the ages you've come a long, long way (evolved) from the understanding that your ancestors held when they first roamed the earth and considered the heavens. As far as they could comprehend, in the beginning the earth was flat and contained vegetation, water, animals, the earth below and the heavens above with light and rain. In Faith they declared their conviction that "someone must have been responsible for creating all this", and so they recorded *"In the beginning, God".*

Today you stand in different circumstances with the benefit of microscopes that can see with incredible accuracy the amazing depth of the natural / physical universe. You now are able to probe the depth of the smallest particle of physical reality, the Atom, and you also probe outward through powerful telescopes and are overwhelmed by the majesty of the heavens filled with distant mysteries of the universe. Microscopes and telescopes have greatly expanded the extent of your knowledge.

Beware! Just because the extent of your knowledge has been enlarged does not necessarily mean that you have greatly advanced in your understanding of what is essentially true. In reading the following essays I've attempted to help you grasp the unity of who I AM and all that I have created.

Here is your important and challenging task. In the midst of your hyperactive, busy, sound filled, enticing, ever-changing world it is critically important to swim against the current of modern life. You must decide to consciously implement your *ASMize*: ASMize, grow quiet to reflect, contemplate and meditate with a seeking heart and yearning spirit that "goes beyond" the familiar, predominant, rational, intellectual and reasoned thought patterns which you inherited from yesterday. By choice, you must move beyond the intellectual and measurable toward the "inner essence" of the Always Something More (ASM) that lies beyond and within what is normally seen. This requires a time of apartness and quiet. There are no shortcuts. Amidst apartness and the quietness you will catch glimpses of a relationship between secular and sacred, scientific and religious, natural and spiritual, and intuitively gain insight into what the normal senses of sight, sound, smell, taste and touch can contribute when each is interrelated and bonded to the **A**lways **S**omething **M**ore, which is who I Am.

Always Something More

When you go beyond the first rational and emotional impression that rushes immediately into your mind and begin to probe more deeply to capture the essence at the core of what is being considered, you will increasingly discover that you are on a journey, the end of which is not in sight and can never be finally attained! *It is a journey filled with joyful anticipation as you encounter both mystery and paradox*! It is simultaneously playful and serious.

If you have the patience to truly seek from the depth of your being, and the subsequent will to apply the intuitive insights that trigger a moment of "aha", you will discover that there is no chasm between them. They're not antagonistic, separate and distinct from one another. Indeed, they coincide and enhance each other. It is my hope that you will increasingly realize that they are one *"in essence"* if you but have eyes to see through a quiet, reflective attitude that enables you to comprehend, and the willingness to apply what you discern in your daily life! Remember, you are on a journey knowing you will never arrive at a final destination because of ASM

In these essays, I am probing *within* the Christian tradition to discern the Spiritual Awareness found in the life and teachings of Jesus who lived over 2000 years ago and died at approximately age 33. The written record of his life encompasses only the last three or four years of his life on the earthly scene, yet, through the written record of what transpired during those three years his Spirit has consistently claimed hearts and souls and transformed lives. It is from the perspective of Jesus that I am writing these essays. I was essentially one with Jesus and expressing my essence through him, for I am God, ASM!

Always Something More

I am using one of your brothers on the human scene to serve as my agent in publishing and distributing what I have written. If you need a name, call him John since I understand that you human beings are always needing to "go to the john". (Yes, I do have a sense of humor!) I desire that you focus on the message rather than the person I have chosen to publish and distribute. I have chosen him because of his Spiritual Awareness as received in the process of his journey as a disciple of Jesus who claimed to be, "The Way, The Truth, and The Life." Remember that Jesus said "do not think that I have come to abolish the law or the prophets; I have not come to abolish them but to fulfill them". Jesus promised that those who genuinely seek to be living instruments of the essence of what he taught and lived would find a Wholeness Producing Spirit (Holy Spirit) that would abide within and speak through a deeply burning heart.

My servant, who is serving as a conduit, makes no claim of being a scientist or nuclear physicist. He is simply an ordinary human being like yourself with availability to "Google" as a search tool and what can readily be learned by anyone who is willing to look at "science as explained to children in our modern age". You can readily check out some of his assumptions on the Internet.

In these essays I am sharing interrelated and interacting terms and concepts that metaphorically point the way toward *a meaningful base of communication to resonate with any human being; theist or atheist, religious or secular; anyone who is willing to be a part of the journey.*

ASM always contains paradox and mystery. In these essays I am attempting to write in a personal, yet non-personal, manner (paradox). I am having my servant copyright these essays only to non-copyright them by *declaring that they are unconditionally available to the largest possible audience; however, whenever, wherever.*

 have instructed my servant not to be concerned about plagiarism which is totally out of place when referring to and utilizing anything contained herein. I reminded him that anything he has received in these essays and anything he choses to do with them is to be freely and unconditionally available to be released. You can copy, quote, publish or whatever. I have instructed him to use a pen name through which you will discover who I really am.

You human beings are always concerned with who gets the credit. You seem to assume that what you call a person's "credentials" automatically communicates truth and insight. You forget that all through Scripture I've chosen ordinary persons to be my channel of love and grace. Credibility had no place in whom I have chosen. None of Jesus disciples had obvious credentials and none were in positions of religious authority or had the stature of outstanding social recognition.

Always remember this! Truth is <u>not</u> dependent upon the credibility of the author or speaker. Truth is always **self authenticating through the process of application***!*

1. Osmosis
Gradually Absorbed and Appropriately Filtered

One of the first realities that nature forces you to consider is that water is crucial for life! All things that you describe as being truly alive contain water. The adult body of a human being is composed of approximately 65% water. But that is not the total story! Consider the importance of osmosis.

Osmosis is the process that absorbs water gradually, and then appropriately filters it so that it can accomplish its intended purpose. For water to truly give life it must go through the *ASMize*: gradually absorb and appropriately filter.

- A sponge does not instantaneously take in water, but rather, limits the intake in order that it may be appropriately retained and distributed.
- The same is true with all plant life. Plants have roots that do not drink water, but rather gradually absorb through tiny holes on the outer layer to filter it and benefit every portion of the plant appropriately.
- The human body can instantly take in a certain amount of water by drinking, but then osmosis immediately begins to take place and the water is absorbed through the outer layer of every cell very gradually and then filtered so that it may appropriately nourish the particular cell and ultimately the body as a whole.
- Our kidneys function by osmosis when they gradually absorb water, and appropriately filter it, eliminating the unused portion as urine.
- Leaves on trees utilize osmosis to convert the rays of the sun, gradually absorbing and filtering-the

rays so they can appropriately nourish the entire plant.

When we observe how nature functions and the message it communicates, one unavoidable and crucial message is unmistakably clear.

Gradual intake and filtration produces healthy results rapid intake without filtration frequently produces unhealthy results

- A gently flowing stream nourishes as it flows, while a rapidly flowing river erodes as it rushes along. That's the reason you have a Grand Canyon.
- Floodwaters are almost always destructive.
- Shoveling food into your mouth in large amounts without chewing and allowing the liquifying and beneficial action of saliva and leaving the appropriate taste buds unsatisfied, is the primary factor in the creation of obesity and an unhealthy body.
- A gentle breeze satisfies and sooths while a hurricane or tornado destroys.

When you ask the question, "what can Nature teach us concerning the essence of life", you discover that Nature clearly reveals that superficial and *too rapid intake* and *not making the effort and allowing the time for filtering* frequently results in destruction and chaos.

The teachings of Jesus clearly reveal this awareness when he stresses that he did not come to eliminate important laws given by Moses, but rather to reveal their depth and essence. To accomplish this he frequently used parables and metaphors where one must search for oneself and then apply an appropriate action to discern what is or is not valid. This is specifically revealed in the insistence of Jesus that he did not come to destroy the law, but rather, to fulfill it and reveal its essence and core. Matt. 5:17-20.

These essays clearly emphasize *the ASMize* again and again by asking you to avoid all quick and automatic response and to choose instead to "grow quiet, breathe deeply and then rhythmically, relax, meditate, and reflect that you may discern the Always Something More present amidst all that exists. Without using this osmosis method the odds are that what you read will not make sense and will never grip your soul.

In the following essays I will be referring to this ASMize by a new term I am creating, "ASMize"?

2. THE BIG BANG the "URGE TO CREATE"

Where shall I start? Why with the Big Bang of course! This is a theory in astronomy that assumes the universe originated over 13 billion years ago in an explosion from a single point of nearly infinite density of energy. Prior to that moment there was nothing; during and after that moment there was something which we describe as the COSMOS, the universe. While most modern scientific research has no way of clearly explaining in detail the what, why, and how of the Big Bang event, there is general agreement in your day that out of the chaos of nothingness a "Big Bang" occurred which projected the universe into existence.

Following is my description of the Big Bang;

The presence of *an immensely forceful energy field, filled with potential that was yet undefined and indescribable, yet motivated toward creation.*

In my essays *I am* summarizing this reality by the phrase "*The Urge To Create*". That is a description that truly reveals the core of who I AM.

My universe, this immense, energy-filled-force-field, has always contained two dynamic realities:

- Infinite Potential --- that is undefined and indescribable.

- And the Urge to Create, a powerful, unrestrainable force that is expressed through an explosion of suffering and, "painful" realities.

On your earthly scene the Urge To Create manifests itself in an infinite number of ways. It can be found in the explosive heat of a volcano, the force of a whirlwind or tornado, the terror of an earthquake radically changing the landscape, or innumerable other manifestations you refer to in nature as "an act of God".

The less dramatic, but nonetheless explosive and painful, expression of the *Urge-To-Create* dimension of the Big Bang, is manifested in the fact that every baby chick must go through the trauma involved in breaking out of its shell. It is manifest in every seed that bursts forth from the security of its container into a new and strange world. It is present in every cry of pain as a mother gives birth and in every human embryo that goes through the trauma of being expelled from the security of its mother's womb.

The *urge to create*, always involves some form of the frightening and confusing unknown that is an integral and essential part of the birthing process. Realistically speaking, each of the above illustrations represents a smaller manifestation of The Big Bang. That this powerful urge is built into all of nature is a self-evident reality.

When you move beyond the material world and consider my ASM perspective in the world of spirit you will find the same revelation. The creation story of Hebrew / Christian Scriptures portrays the entire above-mentioned process through the metaphorical story of the beginning of creation. The Ancients were striving to describe this force field, incorporated in the term "God," that had at its core the incredible Creative Urge that burst forth and kept unfolding in ways filled with creative potential arrived at through difficulty, suffering and hardship.

Always Something More

The creation story incorporated in the book of Genesis includes the creation of Homo sapiens, human beings, who are *specifically created as significantly different from other animals and commanded to continue the creation process on the earthly scene.* Human beings were given instructions that they were to play a dominant and unmistakable role in helping this creative urge to unfold in a wonderful way. This is described by the metaphor of the Garden of Eden which is filled with an infinite variety of wonderful trees containing infinite creative potential.

The story then metaphorically describes how hardship and suffering entered the scene. I will deal with this portion of the story in the following essays. For now, suffice it to be aware of the significant awareness that the Urge To Create contains the necessary and inevitable reality that some form of suffering and pain will always be involved when something new is be born.

I am ASM. I am the one who sent the person, Jesus, who clearly taught you that to follow him and participate in the joy of the ongoing Urge to Create will inevitably include the experience of suffering and painful aspects of the Big Bang in the form of rejection, persecution, hardship, and tribulation as an inevitable part of something new being born. The Ultimate fulfillment of MY intent and purpose happens only through some manifestation of that Big Bang reality.

The journey Jesus made, at his Heavenly Father's insistence, included rejection, pain, and suffering that culminated on a cross at Golgotha and then continued on through the resurrection into the Always Something More of resurrection and an ongoing journey! *Jesus's experience on the road to Calvary is an inevitable, spiritual part of the Big Bang.* It is not a tragedy that could have been avoided. This is mystery and paradox in the extreme! Calvary is the supreme expression of God's infinite love! Jesus insists that his followers also experience pain, rejection and hardship and would need to "take up their cross and follow him. Paradoxically, a part of the *"good news of the Kingdom of God being at hand"* that Jesus proclaimed inevitably involved spiritual dimensions of the Big Bang. Jesus said "In the world you will have tribulation, but be of good cheer for I have overcome the world" John 16:33 (RSV). Infinite and unconditional love always includes self-sacrifice and the bearing of pain. More about that later

Now, ASMize. Take a few deep breaths and let your being grow quiet. Join me as I probe the question of what ASM spiritual insight can be discerned from the Big Bang and how that insight should affect your awareness and subsequent actions.

The *Urge To Create* is present in every human being. Hand a child a crayon and a piece of paper and the resultant wonder of this urge gets eagerly posted on mother's refrigerator door! Yet you so easily forget that this *urge to create* has an explosive dimension that requires that you "pay the inevitable price" if you intend to grow in and through what you are creating. Every form of birthing necessarily includes both pain and gain! We must expect hardship, uncertainty, failure and confusion amidst life because it's a necessary and foundational part of living amidst the Big Bang that is at the core of the universe.

You must willingly learn to accept and face challenges with an awesome respect. Know the frustration of not

easily accomplishing what you are trying to do, and encounter. uncertainty and unpredictability. Some form of suffering and pain are an inevitable part of being truly alive!

The *Urge To Create* always moves beyond the accepted, secure and known into new territory including that which we consider unacceptable, insecure, and terrifyingly unknown. The *Urge To Create* w*ill always include some form of pain as well as gain, some form of dying and rebirth.* Yes, both Nature and Scripture reveal we must learn to get beyond our desired and insistent demand for a safe, secure, comfort zone and face anxiety and fear of the unknown head- on, as we keep moving forward. This is an inevitable part of truly being alive!

The Big Bang unmistakably teaches us that to choose a stagnant, comfortable, totally dependable and predictable lifestyle is to *not truly be alive at all*! At the beginning of life, and through every stage of growth that you experience there will always be some form of Big Bang that projects you into the frightening and risky unknown. Your insistence to always live in secure and comfortable circumstances, free from the unknown and hardship, is to deny the very essence that is at the heart of all reality.

Indeed, most of you have been raised in homes were your parents sought to always keep you safe and secure as they sought to give you all the comforts and desires that life offers. But I ask you a crucial question! When this loving desire is overwhelmingly present, is it truly in touch with the essence of reality? Isn't it equally, or more important, to constantly learn to look at the cost involved in growing and achieving? Doesn't "being alive" involve much more than being comfortable and receiving instantaneous achievement of every desire? Haven't many of you been overprotected and overly provided for without sufficient personal effort or cost? Indeed, doesn't your modern cultural obsession run counter to this awareness of the Big Bang and place the highest priority in life on immediate satisfaction in a comfortable and effortless manner that produces pleasure and happiness? Is it not true that all modern culture and advertising is dramatizing a way of life that is directly contrary to the true essence at the core of the *Urge To Create*, the essence of ASM!

Notice my use of the term "urge". "Urge" refers to a supremely imperative DESIRE. Jesus made perfectly clear to his disciples that "God is a Spirit, and those who would worship God must worship, (*deeply desire*) in spirit and in truth"! The choice is always yours. What do you truly desire at the core of your being and what are you going to do about it? Ask yourself these questions.

- Am I daily striving to go beyond the easy and comfortable and participate in the Big Bang?

- How do my daily choices involve The Creative Urge with the accompanying exciting, enticing, fearful, explosive dimensions that are always present and inevitable?

- What is it in my daily life that I'm trying to avoid and refusing to look at?

- Am I willing to smile amidst pain?

- Am I always grateful for every little blessing that comes my way even as my physical and material resources diminish.

Every stage of life contains a different answer to these questions! The necessity to continue creating still demands expression. The Big Bang is always there if we but have eyes to see, ears to hear, and a true desire to understand and apply. Are you ASMizng; gradually absorbing and then appropriately applying?

3. Orderliness and Similarity

Always Something More

Humbly and gratefully acknowledged

As you reflect upon the results of the Big Bang, the dominant awareness that bursts amidst nature's presence is the manifestation of order and similarity. You gaze at the awe-inspiring patterns in the heavens above and immediately recognize an orderliness that has been present since the beginning of recorded time. Each day the sun arises and subsequently sets in its unfailing, orderly manner. The seasons unfold one after the other: summer and winter, springtime and harvest. The more you observe the animal, vegetable, and mineral dimensions of existence you recognize regular, orderly, instinctive, behavior. Things are as they were meant to be, dependable and orderly.

Simultaneously, you discover that orderliness includes what you would term "similar categories". Trees are trees, birds are birds, water is water, and all of nature has a category that defines its similarity to other related items. With your modern technology you define and measure this orderliness and similarity and develop names for the various categories visually observed: cells, molecules, atoms, elements, families, religions, political governments, sports, etc. are all manifestations of various forms of orderliness and similarity that are inherent in all of creation and the process by which creation continues. It is the basis of modern rational thought and scientific progress, and all human relationships.

In the creation story found in the book of Genesis in the Hebrew / Christian Scriptures you find this orderly development of similar realities spelled out in an unfolding pattern. All of creation is recorded as having happened in a very orderly manner with all forms of similar matter created with specific content. At each stage I (ASM) am described as saying "this is very good". It is the very foundation upon which rational thought is built. You find comfort, security, and hope in the ever-present orderly and similar (categories) that embrace you, surround you and enable you to relate meaningfully and helpfully in the midst of daily life. It is the opposite of chaos and gives you a sense of direction that helps you move forward with the degree of dependability and hopeful expectation.

From a spiritual standpoint the record of "God's Chosen People" unfolds through a man named Moses who goes apart to the top of the mountain to communicate with me, "the source of the highest", and is given tablets of stone on which is written the 10 Commandments which are orderly and similar instructions that can be depended upon to produce the most meaningful and appropriate results. Even prior to the time of Moses historical records of the time reveal sets of rules, regulations, laws, customs, and traditions that clearly manifest the need for order and similarity. It is necessarily so. If it were not so, only chaos would result. You naturally and inevitably understand the need for following the instructions and cooperating with the customary forms and norms society around you dictates.

 Here I come! ASM! There is Always Something More! The spiritual dimension of this orderly process in the book of the creation story includes an event that takes place in what you refer to as "the Garden of Eden". The garden includes an infinite number of wonderful trees, all of which contain fruit to provide sustenance and creative opportunities for humankind (represented in Adam and Eve) to utilize as they serve as co-creators on the earthly scene. However, in the Garden with its many trees there is one tree at the center that is known as *"the Tree of Knowledge of Good and Evil"*. This tree occupying the central position in the garden, represents the reality that the fullest and most comprehensive knowledge of how human beings should produce and create in a healthy manner can ultimately only be known by the Creator. Humankind is warned to "not to eat of the tree of the knowledge of good and evil". To do so is going to create chaos and death.

This metaphorical picture was intended to convey that you are never to assume you have the total answer or complete insight. Humankind is specifically instructed to joyfully, and gratefully continue the creative process **but to do so humbly and gratefully**, and never to arrogantly assume you have the one, true, correct perspective.

As every inventor and creator will tell you, "you must follow the original creator's instructions carefully and not simply follow your own inclinations"! To fail to do so will ultimately result in calamity and chaos. To arrogantly pretend that you are wise enough and have sufficient insight to claim that you have "THE RIGHT answer and perspective", regardless of what the instructions say, is never essentially in tune with the way of life.

The core of unhappiness in the world is a yearning for the right to exercise authority over others and the claim that you alone have the right answer that is appropriate for everyone else as well. The capital "I" at the core is the problem! Every time you assume that you are at the center of what is being considered and what needs to be happening, you are missing the mark. You are refusing to acknowledge that the self-centered, "I" is causing the problem by refusing to stay in touch and interact with the Always Something More that is at the core of all that exists, ASM.

Dangerous territory! Arrogance and closed-minded assumptions based on your limited human understanding transforms orderliness and similarity into a God to be worshiped that denies the reality of ASM. On the one hand you must acknowledge that patterns of orderliness and similarity are crucially important and must not be flippantly ignored. But it is equally true when you do not exhibit humble gratitude and openness, but rather are arrogant and insistent that your particular perspective is the only one to be considered and applied, chaos and destruction are the result

An improper awareness of the place of order and similarity in the midst of creation includes:
- The tendency of some parents who assume their parental role in an insensitive, authoritarian, and dictatorial matter, without loving compassion and mercy.
- It is reflected in the creation of laws that must be applied and interpreted "to the letter of the law" irrespective of surrounding circumstances.
- It is frequently recognized in the tendency of secular and religious groups to engage in ethnic cleansing.

- It is revealed in the existence of political parties creating deadlock because they are determined that they alone are "right".

Now ASMize, *P*robe the question of how this appropriate and necessary need for orderliness and similarity reflects itself in your life patterns? Are you able to recognize ASM amidst your orderliness and similarity so that you also possess open-ended humility and expectation that always allows for something more? Ask yourself:
- Do I respect the necessity of structured, and orderly ways with care and sensitivity?
- Am I acting as though rules do not apply to me?
- Conversely, am I acting as though my particular insight and perspective is the only one that will work?
- Am I so locked into orderliness, and similarity that I am unwilling to have an open mind and probe other possibilities?
- Do I reflect the pattern "I've already made up my mind, don't bother me with the facts?"
- Am I meek and humble enough to set aside what "I" think and feel and truly listen to all aspects of what the other person is saying and attempting to communicate from deep within, and then respond in a caring and loving manner?
- Is ASM truly at the core of who I am and what I'm striving to become?

Orderliness and Similarity, humbly and gratefully acknowledged, is naturally and spiritually *The Way Of Life*.

4. *INFINITE* DIVERSITY

Hidden amidst order and similarity

Always Something More

From the earliest moment when you began to observe, you discovered that, along with orderliness and similarity, existed the wondrous, awe-inspiring presence of *infinite diversity.*

The examples are too numerous to even begin trying to define. They compose the infinite variety of sub-forms within categories: animal, mineral, or vegetable. Different species of trees, birds, butterflies, fish, whatever, appear in unfailing, infinite variety.

Through your modern microscopes you have learned *of the infinite diversity that exists at the very core of creation.* Amidst all the leaves of the forest you will never find two exactly alike. Every snowflake that falls is different from every other snowflake. Even more, every fingerprint on every human being is different from every other human being that has ever lived throughout the course of history. And this diversity can now be measured scientifically with modern instruments through a process known as DNA. Talk about infinite diversity, WOW!

This *infinite diversity* as an essential element within the *Always Something More* was spiritually indicated by Jesus when he said, "every hair on your and is numbered" and "God knows every sparrow that falls". This wondrous diversity and uniqueness has never truly been grasped in its fullest dimension by either science, philosophy, or religion; yet it is an awesome and inherent part of ASM!

TAKE NOTE! A major element creating havoc and confusion in your world today is your determined refusal to acknowledge the inevitability and wonder of *infinite diversity!*

Basic humility and deeper awareness of *infinite diversity* is not allowed its proper place in the functioning of daily life. Instead, you insist that anything different is something less, inadequate and lacking rather than distinctive, unique, and filled with wonder containing unique potential!

Your refusal to celebrate the reality of infinite diversity prevents you from seeing the wonder that is inherently present in all of existence and every other human being!
- It causes you to belittle and demean others who are not like you.
- It causes tribes and nations to create ethnic cleansing.
- It causes political parties to draw hard and unyielding lines and create stalemate.
- It establishes social systems to decide who does or does not belong.
- It causes mocking and bullying toward those who are different.
- It utilizes prisons as opportunities to punish rather than rehabilitate.

I, the Infinitely Diverse One ASM insist that you stop judging and condemning that which you can never fully understand and comprehend! Live with openness, caring and forgiveness that you may not limit the wonder that is occurring everywhere beyond where you can see. That is why Jesus said:

- "Do not judge, and you will not be judged. Do not condemn, and you will not be condemned. Forgive, and you will be forgiven.[38] Give, and it will be given to you, good measure, pressed down, shaken together and running over, will be poured into your lap. For with the measure you use, it will be measured to you." Luke 6:37,38

Always Something More

Stay aware of, rejoice in, and celebrate the infinite diversity that ASM has woven into all of creation and every human being!

Now here is the amazing, truly wonderful part. Are you ready? ASMize! Stop, meditate, and reflect. *Take a deeper look at your own life and celebrate the wonderful unique diversity that is YOU!*

- *No one quite like you has ever been born before on the face of the earth!* Do not let any attempt by anyone else to demean, berate, or bully you because you are different drag you down and keep you from recognizing *who you truly are*. It is their insistence upon commonality and uniformity that is the life-defying and destructive force, not your uniqueness. Every attempt to demean, belittle, or berate is anti-life because it does not allow for the *infinite variety* and *Always Something More* that is at the natural and spiritual core of the universe.
- Acknowledge the wonderful and inevitable *infinite diversity* that is a part of who you are. Others may not consider your difference as a part of their framework as to what is right or proper or good or essential, but that's their problem not yours! You are uniquely and wonderfully YOU. Celebrate that reality, and allow others to have their wonderful diversity as well.
- Grow quiet! Go deep within! Take an honest look at the universe around you! Pay attention to this because it is crucial to grasp and apply.
- Don't focus on the "ain't it awful" dimensions of what is wrong and complain and criticize. *Rather, proclaim and celebrate the infinite diversity that you see.*

- Humbly recognize infinite diversity in yourself and in all others as well as at the core of creation!
- Yes, you do have things in common, but you are also uniquely and infinitely different and wonderful!!!
- Stay open! Stay alert. Look for the small, unique core of what others are feeling and attempting to communicate through the inadequacy of words, rather than jumping to inaccurate assumptions that you know the depth of what they are attempting to communicate. Remember that they are a part of infinite diversity as well. Listen with your heart. ASM.
- Ask yourself the question, do my words, actions and attitudes truly reflect a non-judgmental, not condemning, forgiving, caring, helpful respect for all others?
- Does an awareness of life's infinite diversity result in a determined unconditional love and acceptance of myself at the core of who I am and what I am becoming?
- As I observe the world around me, am I constantly filled with an infinite gratitude and profound respect for the diversity that is always there?

5, AN EVOLUTIONARY CORE

The Restless Energy Of Becoming

Always Something More

I am ASM and as you reflect upon the Big Bang that is at the core of my Urge To Create, you rapidly begin to realize that this inherent, energy-filled Creative Urge containing orderliness and similarity along with infinite diversity is also filled with *"the restless energy of becoming"*. This takes the form of *a forward thrust that is filled with endless potential.* At the core of all nature in its many manifestations there is yet the unformed and unknown reality that is yet to *evolve*. It is a process that is inherently filled with potential beyond what exists and eternally moves outward towards something more that has not yet been defined. Everything must continually evolve!

You refer to this *"restless energy of becoming"* by the term **evolution**. Ever since the beginning, the universe has been thrust forward into ASM… Always *Something More.* The Big Bang reveals this dimension of its reality in one way or another in every resultant form of nature. It is manifested in all things: animal, vegetable, or mineral. This evolutionary thrust is mysteriously and paradoxically both awe-inspiring and frightening, desired and unpredictable. The ultimate importance and understanding of what is happening is unknown and scary, yet the need to move forward is inherently present.

While Charles Darwin is the person highlighting one aspect of how this evolutionary process manifests itself, there is so much more (ASM) beyond what Darwin is attempting to describe. Archaeologists are continually studying and measuring how this has manifested itself in ages gone by. By carbon testing they can detect how the process has worked and what the results and manifestations have been. Chemists are continually researching ways to evolve new medications beyond those of which we have been previously aware.

Always Something More

Cosmologists tell us that the universe itself, with its billions of galaxies, is endlessly expanding. The cosmos is not static and fixed. From the standpoint of the First Bible that ever existed, which we refer to as Mother Nature, we discover the wonderful and fearful, enticing and threatening fact that EVOLUTION is at the core of all that exists. The universe is continually expanding! I, ASM, am by no means finished creating. It would be contrary to who I really am.

Even so in the realm of Spirit. In the creation story in Genesis, the powerful reality of *Evolution* is revealed in the process of how creation occurred. Creation is metaphorically described as an evolutionary process wherein creation takes place "day by day". Intriguingly, along with this evolutionary process, the phrase, "saw that it was good" is frequently expressed by God, in communication with the heavenly host, as the process takes place. Your ancestors who repeated this story were really onto something!

Subsequently, the creation story reveals the necessity for human beings to continue evolving by implementing their creative urge in a manner consistent with what I intended, in my image.

Change, evolution and growth are unavoidable and profoundly at the core of ASM. You either evolve and grow or you die!

> *To remain static is totally contrary to nature and you must never be rigid and totally resistant to change.*

Stay alert! I want you to know something very special about this evolutionary process. In nature this process includes the possibility of *mutation*, which is the basic change within all living things that enables them to always adjust and move forward according to the circumstances at hand. All of nature, internally and externally, knows it cannot remain static. It must adjust to truly be alive.

Now it is time to stop thinking and analyzing and ASMize, reflecting for deeper insights. Become quiet, relax, breathe slowly and become open-minded in a state of receptive wonder. Continue to probe what is the essence at the depth of evolution both naturally and spiritually.

When you consider the varied insights and arguments that have been taking place regarding creation as recorded in the book of Genesis, you soon discern the tragedy that some non-evolving, rigid and static forms of thinking have ignored the evolutionary dimension.

Many of you consider yourself to be modern and intelligent persons who are determined to be scientifically accurate, but in the process you have diverted attention from the crucial awareness of evolution. Looking at the creation story you argue about seven days being literally necessary and what was created in what order. You declare that everything that was created literally happened exactly as recorded. When you return to Scripture humbly and with an open and receptive mind and heart, you discover the core reality that goes far beyond your petty, presumptive, analyzing ways. Scripture envisions that you human beings are going to continue evolving as you relate with the essence of all that exists which is evolving as well, even the Scriptures whose authority you rightfully proclaim.

The goal of your reflection is to discover *ways you need*

to change your attitudes and actions to truly interact with the evolutionary essence of all creation and core spiritual values. You need to join me, ASM, and personally ask some of the following questions:

- Am I so locked into traditional ways of living, thinking and acting that I am daily refusing to *evolve into the forward thrust* that is the essence of nature all around me and an inherent part of who I am truly meant to be?
- Am I truly looking for the ASM that is unfolding in front of me <u>each day</u> in all of its ASM , Big Bang, infinitely diverse wonder?
- Am I wishing things could be as they were yesterday instead of facing the challenges of today with daring, courage, and hope?
- In the past I have been focused on self-congratulation for achievements, and self-blame for failures. Am I moving forward beyond all this self-centered action?
- Am I choosing instead to look forward to the energizing work of the Wholeness Producing Spirit that is present within me and beckoning me onward?

Again we discover that the natural world and the world of spirit have a message that is essentially the same. Because each of us is infinitely unique and diverse, this evolutionary unfolding is always going to occur in a variety of ways. It will always be manifested differently in each of us at different stages of our journey. But the basic truth remains. We must not stagnate or spend our time fruitlessly looking backward. Our focus must always be on the future. You are inherently filled with

the restless energy of becoming.

6. The ATOM

"Energy In Purposeful Motion".

Always Something More

Think "Personality"

To probe something until you discover its core reality is to discover its "essence". Utilizing insights received through modern technology, let us continue to probe ASM, the "awesome" essence of the unknown and forceful energy that created the Big Bang. HOW? By continuing to examine the evolving results of the Big Bang, we can discern innumerable new dimensions within matter itself.

You are surrounded by the physical universe which, in the science of physics, is given the term "Mass". The form of matter or Mass in question may be animal, vegetable, or mineral. It matters not which form is chosen. When you analyze physical matter in ever smaller dimensions you come at last to the smallest measurable form which is the *"Atom"*. In your modern educational system every child is taught from the earliest age that the *atom is the basic building block for all matter in the universe.*

Within the atom are the following elements: protons, neutrons, and electrons. Protons and neutrons are within the nucleus of an atom while the electron orbits around the outer edge. A single atom is not matter but rather a combination of elements in purposeful motion.

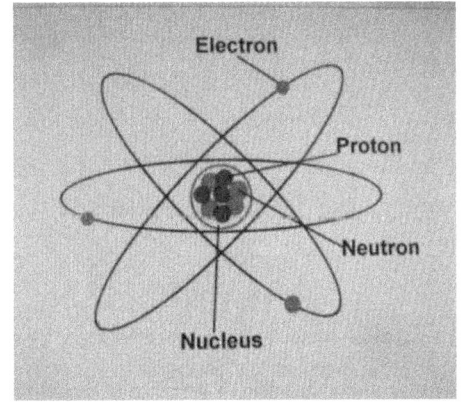

It is the capacity of atoms to combine that produces matter.

Always Something More

Pause and grow quiet. Think about this statement and let it permeate deep within. Nothing is truly solid at its core.

All matter in the universe is ultimately a combination of "bundles of energetic elements in purposeful motion toward a common objective."

Each bundle of energy is continually interacting with the others and fulfilling its appropriate role in relation to the WHOLE!

Within the realm of quantum physics there is a specialty that examines sub-atomic elements, and this scientific effort is continually seeking to discover the ASM (Always Something More) within these energetic, always in motion, elements. A recent discovery set the scientific world celebrating!

Amidst these elements within each atom there exists a previously undefined component whose purpose is to see that every other element within the atom functions with respect and purpose, interacting with and harmoniously caring for every other part. This previously undefined reality was hinted at by a number of physicists, but the person chosen to have his name forever linked with the discovery of this element is a scientist by the name of Higgs. The existence of this sub-atomic component was only a theory until it was finally verified. It is known as "Higgs Boson".

The reality of the Higgs Boson component was finally proven and acknowledged in 2012. A globally coordinated effort of over 110 nations at a cost of over 10 billion dollars, resulted in the creation of the HADRON COLLIDOR ACCELERATOR, the largest structure ever built. It is a 17 mile circular tunnel under the city of

Always Something More

Cern, Switzerland including a portion of France. It is a continuing effort involving over 1000 scientists and technicians across the world. Its purpose: to study and analyze super, sub-atomic particles.

On July 4, 2012, scientists conducted an experiment utilizing the Hadron Collidor to confirm the existence of the Higgs Boson.

Be patient. Hang on tight. You certainly cannot process and grasp all this intellectually, but as we journey together you will begin to grasp its hidden meaning beyond that which is observable through sheer technical terminology. For now, simply know that modern quantum physics includes the acknowledgment of a component that is:

<u>*relational and interactive, and enables all of the other elements within the atom to relate to one another in a meaningful, synchronized, harmonious, purposeful and mutually respectful manner.*</u>

If we were to create a general statement concerning the *essence* of this component, it would be fair to say that it enables the atom as a whole to function in a meaningful, synchronized, harmonious, and, purposeful manner. Some scientists have even dared to refer to it as "the God particle".

Now, *ASMize*. When you humans try to describe the nature of another person you say that you are describing his or her personality. P*ersonality* is a combination of Something More that goes beyond mere physical existence. It includes how you relate, interact, and choose what is meaningful and move toward a given purpose which is the essential totality of who you are, internally and externally. It includes how you relate to all that is about you, As a human being you would be described as a person with a particular "personality".

Hang on to that statement! Ponder it. This description of the governing core of the very center of every atom in existence is identical to the elements you declare to be present in human beings who are persons and have a personality all their own. At the core of each of the trillions of cells within your bodies, there exists this incredibly wonderful component: *personality*.

You unconsciously assume a person's personality is integrated when everything functions "normally" in an appointed, natural way. Each "personalized" atom is intrinsically in touch with all that has gone before, infinitely complex but wonderfully synchronized and "on target" and always evolving toward an ever-expanding future. What quantum physics refers to as Higgs Boson is this core component possessing all these qualities and giving the atom its essential composition, it's personality.

Keep reflecting! Grow quiet and let your heart and intuition speak! Is not this the precise awareness and understanding that religious writings have always attempted to proclaim? Scripture clearly declares that what exists came into being in a purposeful and evolving manner and in a very *personalized* way? That is to say *totally relational, and continually interacting in a meaningful, harmonious, purposeful, synchronized manner.*

This was the instinctive basis behind early human attempts to describe the indescribable by seeing different aspects of nature as containing aspects of who I AM, God. Human beings have always recognized a dimension of the

Always Something More

"God" force in sun, stars, every element of nature, fertility, etc. You h*umans have always sought some way to relate to these larger, relational and interactive dimensions of reality.* Many early American Indian tribes had sacred mountains and gathering places and sought to honor the wonder at the heart of all creation in their attitude toward animals. There has always been an awareness and whispers of a *personalized* Something More.

In the creation story, I gave you humans the co-creative responsibility to continue the creative process on the earthly scene through the declaration that you were created *"in the image of God"*. You were given the privilege of staying in touch with, and relating to, the Creating Force! An image is always a reflection of Something More that is real. When you cease striving to interpret the Bible in a literal fashion, and begin to see it as a wonderful gift and revelation from Me, ASM, God, that helps you stay in touch in a meaningful way with the essence that is inscribed into all of creation, you end up with wonder upon wonder and an ever-expanding awareness.

Stay alert! This instinctive awareness is what humbles you, and causes you to "pray"; which is your attempt to personally interact with this larger personal dimension of reality. This vital energized central response exists in each of the trillions of cells in your body and all of existence, and it is automatically a part of all reality whether you acknowledge it or not. It is *totally relational; continually interacting, meaningful, harmonious, purposeful, synchronized, and moving forward toward an ever-expanding, evolving future.* A you some! ASM!

Always Something More

It is accurate to say that it is an expression of infinite and unconditional love. The writer of the Gospel of John tells us that God is love. No wonder you intuitively and naturally desire to communicate with and interact with ASM / GOD!! It was how you are intended to function at the core of your being WOW! That is truly the essence of what it means to be created in the "Image of God"!

So the crucial question for each of you at every stage of life is the same!

- Are you personally seeking to humbly be aware, sensitive, interacting, and expressing a caring love toward everything that is going on around you?

- How can you more meaningfully reflect the *image, "personality"* of ASM ... the great "I-am-and-I-am always-becoming"?

- Are you finding some way to communicate, interact, and relate to this larger personality intrinsically in the universe in a manner that daily helps you to grow?

- If you don't pray in a religious sense, in what way are you choosing to intimately relate to everything around you?

This is not an intellectual game you are playing! It is a point of reference that enables you to identify and interact with the core of all creation and the natural world in which you live. Are you going to ignore Me or are you going to attempt to find some way to continually interact and join Meyou in the journey? I am ASM. I desire nothing more than to help you become the fullness of all that you are meant to be.

7. ELECTROMAGNETISM

An Attractive Or Repulsive Force

The Necessary Existence Of Opposites,

Most persons are fascinated by magnets. Electromagnetism is a force field that contains *two opposing signs of charge, one positive and one negative.*

Now hang on tight because here is an important and critical reality!

At the heart of each atom and at the core of all existence there is the necessary polarity of both positive and negative force fields found in the elements of Proton (positive, attractive charge) and Electron (negative and repulsive charge). Life cannot exist without this inherent polarity.

You find it expressed in the polarity of light and darkness, night and day, male and female, life and death, sunshine and rain, waking and sleeping, conscious and unconscious, healthy and diseased, -- the list goes on and on! *Without the simultaneous presence of attracting and repulsive forces including a tension between opposite poles, life cannot exist.* It simply is "a fact of life".

You cannot possibly know the fullness of "why it is so" because you are not God. Nevertheless, affirm that polarity *is* an unmistakable necessity at the core of all life. If one pole or the other is significantly overloaded, life becomes unbalanced and distorted. Both poles exist and must be present in a co-existing and helpful manner.

Within the human body a force has been established known as "*the immune system*" which is the presence of cells within your blood continually at work maintaining a balance between the opposite forces that you refer to as health or disease. **The immune system is the body's defense against infectious organisms and other invaders. Through a series of steps called the immune response, the immune system attacks organisms and substances that invade, destroy, and cause disease. When antigens (foreign substances that invade the body) are detected, several types of cells, called antibodies, work together to recognize them and respond in a healing manner so that life can once again become balanced.**

Now think of your personal life as a human being. Amidst your genetic and cultural givens that are unique for each of you, there exists attractive or repulsive forces. To be truly alive is to be choosing which force will prevail in every area of life and this is what creates our personality. I will be discussing this more in a later essay. For now simply acknowledge that by your personal interest and determination of will, you are continually *repulsing* (filtering out) some possibilities by being disinterested, refusing to acknowledge their existence, or actually fighting them head-on. Conversely you are *attracting* other force fields by your interests and desires.

Now pay attention because the following insight is critical!

You quite often ignore ASM, forget that there is Always Something More, and make the fallacious assumption that pain is a negative and destructive force and pleasure is a positive and constructive force. This is a false dichotomy! Pain is also beneficial and good since it alerts us to what needs attention. Too much pain is not beneficial. So you are continually faced with the challenge of how much pain to creatively endure as you seek to return to balance. Remember what I said previously: pain, suffering, and

frustration are always a necessary part of the evolving, process of birth and newness of life?
This is likewise true of pleasure. Pleasure is not always good. When it takes an excessive manifestation and throws us off balance it becomes obsessive and you place too large a value on comfort and ease. You are back to the subject of paradox and mystery that are at the heart of ASM.

For instance, doctors prescribe certain pills to alleviate pain, which is good, but an excessive use of these pills not only reduces pain but also creates pleasure and subsequently becomes addictive and is at the core of one of your biggest challenges; the utilization of heroine which can ultimately kill us.

The more you *ASMize*, the more you will recognize the essential *inevitability and necessity* of opposing forces of attraction and repulsion as a necessary part of life. What you refer to as "evil forces" frequently can affect you in a positive way that helps you to grow. There is Always Something More. Which is to say you frequently learn and grow more from your pain and difficulties that from your moments of comfort and ease. It is in the appropriate balance of the tension between opposites that ASM is found. Yes, I'm there! I am always there.

Maybe you should not be asking the unanswerable question "*why* does so much evil and suffering exist in the midst of my life or in the world at large"?
Maybe the only realistic question is, "since this is what is attracting or repelling me at this moment in my life, what can I choose to do that will help me and my world evolve more helpfully into what I truly would like it to be!"

Always Something More

One reality is unmistakable. At the core of creation polarity exists! The reality is to look at the world around you as appropriately and essentially containing both known and unknown, attractive and repulsive, good and bad, healthy and unhealthy, productive and unproductive forces both of which help you evolve. It's both-and not either-or. Both forces are crucially present and essential. You need to ask yourself:

- What is it that attracts me and why? Will it help or hinder me?

- What things repulse me the most? And why? How can this repulsive experience help me evolve?

Don't spend an undue amount of time on the question of why it is so! Rather, you need to acknowledge that good and bad, right and wrong, helpful and not helpful, will always exist and frequently you will not be able to fully understand why; but you must always face whatever is occurring head-on and-decide what to do about the next step forward. The choice is always yours!

8. HOMO SAPIENS

Creative Potential and the Necessity of Choice

You are asking the question, "do nature and science as rationally explored through natural and quantum physics have anything in common with Spiritual Awareness as explored through religious / spiritual teachings? The pilgrimage continues.

Let us consider the greatest and most miraculous creation of all: a totally unique and distinct creature …. Homo Sapiens. What is it that makes you so totally unique and different from all the rest of creation?

Philosophy has long posited observations on what it means to be Homo Sapiens / "human beings". Much of the observation centers around your minds and brains; your ability to think. But animals, birds and fish also have brains. Their brains seem to function primarily through a process you refer to as "instinctive". Indeed, there are indications that their brains have the capacity to function beyond instinct. For instance, have you ever tried to outwit a squirrel trying to get into your bird feeder? They truly are ingenious! Their ability to discern and adapt is most assuredly not exclusively an "instinctive" process. Animals unmistakably have the capacity to process what they experience and "more than instinctively" adapt. just as you do. What is it, then, that makes us unmistakably different?

You have two distinctive features:

Creative Potential and *The Necessity of Choice*.

Always Something More

I am ASM. You are like me. You contain the creative urge in a way that exists in no other creature. No matter what your personal period of history, culture or environmental circumstance, you are continually called to be creatively involved amidst all the options that confront you and to personally choose amongst the varied possibilities, what you will apply and how you will do it. More than any other creature in all of creation you are distinctly and uniquely, a human being. *You have creative potential on the one hand and the freedom and necessity to choose on the other.*

No wonder the creation story in Hebrew/Christian scriptures includes the creation of human beings, both male and female, as the Crowning Jewel of the Creative Process --- entrusted on the earthly scene with the responsibility of continuing the evolutionary, diverse, creative process by choosing to act in a manner that is relational, respectful, interactive, interconnected, and harmonious (i.e. "in God's Image").

Continue to join me in the journey, *ASMize,* grow quiet, slow down your breathing process, and seek to go beyond your normal patterns of awareness to reflect and meditate.

 Do you remember my observation in the previous essay on Polarity and the necessary Tension of Opposites that is built into all of creation? Inevitably as you daily evolve into the Always Something More (ASM) of who you uniquely are and what you are in the process of becoming, you are continually faced with *freedom to create and the necessity of choice!*

You're constantly aware that in choosing you are not always "on target" (hitting the "bull's-eye") but much more commonly missing the target entirely, even with your best intentions. You find yourself acting and reacting in ways that are not helpful to either yourself or the world around you. The term "sin" in Scripture literally means to "miss the mark". Through the story of Adam and Eve in the Garden of Eden in the book of Genesis, humans are revealed as having chosen to focus on what they wanted to do rather than submit themselves to the intent of the original creator and the needs of the world around them. That is the meaning of the term "sin" (to" miss the mark) in Scripture. Namely, focusing on what you desire rather than what ASM yearns for and desires as best for creation at large. You'd rather capitalize your "I" rather than see yourself as a servant of all.

As you become aware of who you truly are, you have times of infinite gratitude and rejoicing and also times filled with feelings of guilt and inadequacy. This is inevitable. You human beings face the daily necessity of choosing between polarities, "opposites in tension", which includes the awareness that you are going to inevitably have contradictory feelings of either moving forward or falling backward. Paradoxically, you humans are created in such a way that you are going to feel inadequate and frequently guilty, and aware that you are missing the mark most of the time! (Theologians refer to this as "Original Sin"). Your only option is to learn from your personal experience of what does or does not work. It usually takes many shortcomings and failures before you begin to discover something that works. Remember evolution! The target is always in motion which means it is especially hard to hit. *The trick is never to quit but to always learn from your failures and move in the right direction!*

Always Something More

Think about that! I, ASM, GOD, am not a scorekeeper keeping a record of what you are doing, placing everything in a column of success or failure. Rather, I AM a loving father that is always healing, comforting, and providing every possible resource for you to move into ASM…. a wonder-filled and ever-expanding Something More! In spiritual terms, sin is always present but forgiveness and "infinite love and grace is who I AM, always healing, reconciling and restoring you and leading you onward.

Now note! This is crucial! Pay attention!

Your task in life is **not** to endlessly grieve over your "evil", unsatisfying and agony-producing choices and inadequacies and then seek to make endless excuses, or blame others because of negative circumstances and disruptive results. Rather, like nature itself, you have the responsibility to *immediately* activate your "immune system" (antibodies) to supply the corrective that is needed to bring circumstances back into balance. Spiritually, that means to cease rehearsing what went wrong and choose instead to faithfully look forward to what needs to happen in the future. The *immune response is to focus forward with caring and love. Accept God's grace by asking for forgiveness and in turn choosing to be forgiving. It's crucial to the "immune system corrective".* "I'm sorry" and "let's see what we can do to make this work correctly" is really the only evolving, fruitful, and productive focus that works.

Note! You spend far too much time grieving, regretting, and blaming. Then you immediately seek to enact some form of punishment: "I'll sue you", "I demand my rights" "Justice must be served", "punishment is due", "you'll see, I'm going to get you"! Retaliation and vengeance are completely contrary to what nature strives to do! They are not the way of life. Their ultimate outcome is agony, darkness, and continual suffering.

Nature recognizes the malfunction (pain) and then *immediately sets in motion an appropriate corrective measure* (Immune System Reaction) that helpfully removes or surrounds and neutralizes an irritant. When you have an abrasion of the skin the white corpuscles immediately come to the front and begin the healing process. This results in a temporary scab and sometimes a scar. The entire goal is always one of reconciling and restoring to the greatest possible measure of wholeness.

- Consider the Pearl inside of an oyster. A piece of sand or other irritant enters the body of the pearl and it immediately begins to surround it with a substance that eventually causes it to become a pearl.

- Consider the injury on a tree. Some healing force is immediately set in motion, occasionally producing a burl. It becomes a very distinctive part of the tree. Woodworkers treasure the burl and transform it into something incredibly unique and wonderful.

Now switch gears and *ASMize*. Notice my ASM behind Jesus parable of the wheat and the weeds… "let them grow together until the harvest and then God will decide what deserves to be eliminated and what remains." Matt. 13:30. A variety of attractive and repulsive options are continually tugging at the core of your being. Simply focusing on the negative and repulsive and seeking to regret, blame, retaliate, intimidate and destroy is not the answer! Rather, enact infinite unconditional love in the form of tiny positive insights and modifications to promote healing and reconciliation toward ASM.

In the Gospel of Mark, Jesus is recorded as saying "Truly I tell you, all sins will be forgiven the sons of men, and whatever blasphemies they utter; but whoever blasphemes against the Wholeness Producing Spirit never has forgiveness, but is guilty of an eternal sin." This means that to close the door to the inner nudges of the Wholeness Producing Spirit and choose to walk along paths that are "not the way of life" is to create a situation for which there is no cure. It's like being aware of the presence of light and then walking away from it. The farther away you walk the more terrible it becomes until you are finally engulfed in total darkness and despair. The human Christian scriptures describe the situation as hell, "outer darkness where there is weeping and gnashing of teeth".

Here Jesus is essentially saying, *"always forgive yourself and others!"* Study the parables and teachings concerning forgiveness in the New Testament and you'll find that it is without exception the dominant urgent requirement of the choices you make as human beings. You are all bound to fail and miss the mark. It's an inevitable aspect of the polarity and the tension between opposites that is a part of all of creation. The real question is, "what are you going to do next"? The Creative Urge is within you and the necessity to choose stands before you:

- Always be creatively open-minded and responsive to God's nudges and insights revealed in the life and teachings of Jesus and led by the Wholeness Producing Spirit that Jesus places within you.

- Endlessly and always forgive *yourself* and others and keep moving forward as best you are able in the moment at hand.

- Choose outward expressions of words and actions that include asking for forgiveness, acknowledging your part in the misunderstanding, saying "I'm sorry", and giving forgiveness with genuine caring and love!

- Unavoidable! Absolutely necessary! Balance and wholeness cannot be restored without it. Reconciliatory action is necessary!

Meditate and contemplate and reflect for there is Always Something More (ASM) endlessly evolving if you are sufficiently humble and receptive to join in the process and grow through the experience. Don't be locked into past mistakes or inadequacies. Never settle for regret, bitterness, retaliation, and endless anger. Forgive yourself and others… ALWAYS.

When the disciples asked Jesus, "How many times shall I forgive, 7 times?" He replied, "70 times 7" which is to say… keep on forgiving yourself and others and keep moving forward. Forgive and keep moving forward as best you are able. Little by little, one step at a time, for I, ASM, am eternally present within you and all of creation evolving the infinitely wonderful Something More that is the caring, interacting, harmonizing, synchronizing, immunity producing life force at the heart of the journey.

You may have been badly hurt and immediately tend to rebel and say, "no way I can forgive". Jesus replies, "I am the one who personally bore and absorbed the worst possible hurt rejection and misrepresentation on Calvary and took the ugliest symbol of hateful anguish, rejection and despair… the Cross on Calvary… and through my self-sacrificing love, I have transformed the cross into the greatest symbol of hope the world has ever known! With me guiding and enabling you through my Wholeness Producing Spirit, nothing is impossible! I AM ASM! By my grace, you can and you must!"

9. Gravity

Pulling Everything To The Center

Gravity is a force pulling all matter (which is anything you can physically touch) toward a dynamic central core. The more matter, the more gravity. Substances that have a lot of matter, such as planets and moons and stars, pull more strongly. Gravity, or gravitation, is one of the fundamental forces of the universe. In everyday use, it describes the force which causes objects to be pulled toward the center of the earth. Even more, Newton's laws discuss how gravity keeps the Solar System together and Einstein's theory of general relativity is about the role of gravity in the universe.

- The reason things stay on the Earth's surface is because of the gravitational pull toward the Earth's center which has much more mass and therefore exerts a greater pull.

- Gravity guides the growth of plants and other vegetation.

- Gravity makes stars burn by squeezing their matter together.

- The sun's gravity keeps the planet orbiting the sun.

- The further you move away from an object the gravitational force decreases. Astronauts feel weightless and float around in outer space.

- This phenomenon is responsible for the rise and fall of the tides.

Always Something More

In summary, Gravity is part of the essential, universal force that holds all matter together and pulls it toward the center. On the earthly scene Gravity enables us to feel "grounded".

Now is time for your *ASMize*. Pause, grow quiet, breathe deeply and then rhythmically, and begin to reflect and contemplate on the "essence" of gravity in the reality of spiritual awareness.

When you are being "pulled toward" something that is more powerful than yourself you are "gravitating towards". Frequently and paradoxically you have no idea why it is so. There is a mystery about it that you cannot explain but you realize it is unmistakably there.

- If it is exceptionally beautiful or harmonious you say you are "spellbound".

- Some of you are drawn to musical interests and love to play instruments or sing.

- Some of you are athletically inclined and drawn toward sports.

- Some of you are intellectually astute and forever reading, thinking, and analyzing.

- If some object or desire has an exceptional gravitational pull you say you are obsessed.

- Because of your uniqueness you find yourself *gravitating toward* different interests or different sexual preferences.

- Gravity is a part of the attractive impulse mentioned under polarity.

Always Something More

If you are allowing your being to be filled with ASM, you are mysteriously and wonderfully being pulled toward the ASM (Always Something More) that is at the core of all that exists. You are celebrating the unique diversity that is paradoxically and mysteriously at the core of your personal ASM.

You may not have any idea of why you feel as you do; why you are drawn in a certain direction. You only recognize that there is a powerful tug at the core of your being that demands to be recognized and expressed in one way or another. It simply won't let you go. You know that you can't ignore it. It somehow represents the unique and wonderful essence that makes you uniquely and wonderfully connected to something bigger than you are.

You are free to choose how you will respond but you are not free to choose to ignore it.

- What is it that you gravitate towards?

- What are the things that "really turn you on"?

- Is there something tugging at the core of your being that you have been fighting and refusing to let claim as the center of who you really are and yearn to be with your distinct uniqueness?

- Even more important, are you truly seeking the ASM within; the obsessional pull that simply won't let you go?

- Is there a divine dissatisfaction within you that is tugging at the core of your being?

- BUT NOTE THIS! Because you are human and filled with the necessity of choice, it is equally possible to gravitate towards that which is unhealthy, destructive, and evil as toward that which is healthy helpful constructive and a blessing. Which way are you gravitating? Necessarily and inevitably the choice is always yours.

This entire series of essays… this mutual Journey… means nothing if it does not result in new patterns of discovery and action. You dare not ignore the gravitational pull at the core of all creation including the core of your being. Because of polarity a variety of gravitational pulls exist. As so beautifully expressed by St Francis of Assisi,

"Heavenly Father, you have made us for yourself and our hearts are restless until we find ourselves in you"

10. PROCREATION - HUMAN SEXUALITY

Procreation is that demanding, automatic process that enables multiplication and reproduction to occur. It is an intimate part of the continuing Big Bang. There is no way to eliminate the desire and ability to replicate. It is an essential part of ASM, and demands to be expressed, not ignored.

- Procreation is found in the trillions of cells within the human body that have the capacity to replicate themselves and continually do so.

- Procreative action is the process which instinctively enables birds to mate and create eggs.

- Procreation is the urgent process whereby an inner instinct requires all animals to mate and replicate.

- Procreation is evident in all forms of vegetation, and is memorably visualized when a dandelion's yellow center matures into a puff ball. The white fuzz is attached to the seed and acts as a parachute. The wind catches it and can carry the seed great distances.

Homo sapiens have always had an inner necessity to procreate. It exists amidst peace or war, rich or poor. It is the energy force behind what you call sexuality.

Everyone should be taught the natural physical process of procreation. The ovaries in the female contain up to 400 eggs present at the time of birth and they slowly mature, beginning at puberty, and are distributed singularly through a process called *ovulation*, once every 28 days during a woman's child-bearing years. The testes in males do not contain semen and sperm at birth, but rather, begin to produce both at the time of puberty. The resulting semen and creative sperm are very active and possess the ability to fertilize a woman's egg. The process continues throughout most of a male's lifetime.

When one looks at the entire procreative energy process, one begins to discover something significant and different taking place in human beings! The procreative urge is not limited to a specific period of time when the possibility of creating a fetus is present. In humans the dynamic activity that I, ASM, have created behind procreation is intricately so extensive that it is continuous and relates to every cell in the body and every nerve ending and the entire blood distribution system. It is a highly pleasurable process that activates the electromagnetic system within the body creating an increased blood flow, expansion and retraction of muscles, and an increased flow of electrical impulses reaching out to every nerve ending throughout the body creating a wondrous, pleasurable feeling referred to as "an orgasm". It is uniquely interrelated to human feelings and emotions, as well as taste, smell, touch, hearing, and sight and mysteriously, paradoxically, and intimately linked to the heart of who I Am.

So much for a brief statement of the natural, scientific dimension. Now it's time to ASM*ize*, Always Something More. Take some deep breaths, relax, and permit yourself to reflect and meditate. Remember you are dealing with paradox and mystery in the extreme! There is no force field in the human body that is so filled with creative potential on the one hand and the urgent need for implementing polarity issues (the appropriate tension between attractive and repulsive desires) previously mentioned and the distinctly human capacity to combine the creative urge with the necessity of choice!

Always Something More

I share now as the human agent ASM has chosen to channel these essays, I acknowledge that one of the most significant challenges of my personal life has been to wrestle-through to a measure of gut-level understanding of this dimension of life. Exceeding life expectancy by more than 15 years, I still end up referring to this area of life as a mystery because of the scope of its expression in us human beings.

The natural, physical part of the equation is quite clear and available to anyone who is acquainted with the Internet and willing to ask questions of Google. Personal discovery and the possibility of communication regarding this profoundly intimate part of life is readily present. That it is a physical phenomena that includes a process which builds up special bodily fluids, causes the heart to race, varied desires to demand expression and pleasures to be experienced, is self evident. It is a part of the procreative process that should joyfully and readily be acknowledged. It will forever be one of the most powerful driving forces inherent in every human being.

The spiritual part of the equation is much more mysterious and complicated. Like all humans, I have struggled with varied concepts of right and wrong super-imposed by society around me. This is clearly revealed in various terms relating to sexuality. Terms such as: necking, petting, self-gratification, intercourse, making out, loving, masturbation, oral sex, rape, sodomy, romantic love, a wife's duty or a husband's duty, nymphomaniac, the world's oldest profession, frigid, machismo, sexy, impotent, sexual harassment, virgin, playing around, infidelity, fidelity, gay, lesbian, bisexual, transgender, dirty jokes, modern freedom, birth-control, safe-sex, abstinence and a host of "street-language terms .... all stimulate the mind's eye to visualize a wide range of sexual realities that daily impact us as human beings. The varied experiences behind each of these words reveals the infinite array of emotional and spiritual interactions that can and do occur in the mystery of what we term "sexual expression" and cannot be ignored but must be dealt with in some way by each of us as a distinctly unique and wonderful person with our own personality.

After a life-time of reading, living, experiencing, and serving as a personal confidant for persons involved in almost every conceivable situation, I now want to try to formulate some important insights into this ultimately unfathomable mystery called "Human Sexuality". Volumes could be written', but for now, suffice it to say that following are some of my life-enhancing and life-determining conclusions as ASM has been at work within me:

- *We are created to be more than animals acting on instinct. Sexual intimacy is a continuous process for us human beings, it is intended for more than procreation.* For us human beings sexual expression is involved in giving and receiving affection and always needs to be a profound way of communicating tenderness, caring, compassion and the desire to be present and available with a deep level of commitment. Sex was never intended to be experienced in a reckless, flippant, demeaning, brutal, indifferent manner!

- *Human sexuality is also a 'bonding' force*; sometimes the expression of a desire and bond that already exists, and at other times possessing the capacity, properly expressed, to create a bonding that does not previously exist. Bonding is always a part of affection and sexual expression as intended by ASM. *Caring and commitment are at its core*, "for better or for worse, richer or poorer, in sickness or health, as long as we both shall live".

Always Something More

- *Sexual expression reveals the yearning to know and be known at life's deepest and most intimate level.* Sexual intercourse is intended as a source of total inner knowing and sharing between two persons who are committed to unconditional and enduring love'. It is not a plaything to be expressed in a superficial and demeaning or harmful manner.

- *Sexual expression has the capacity to produce profound feelings of physical pleasure, and a sense of total belonging and momentary release amidst life's , frustration, pain, and confusion.* Beyond Procreation aimed at creating a fetus, sexual expression is a process which provides a release from built up stress and pressure. The aftermath of appropriate sexual expression is a peaceful sense of belonging.

- *Desiring the highest.* Deep within, each of us resides the necessity of seeking the ASM that is at the core of sexuality. ASM reveals the infinite diversity and uniqueness at the core of each of us, necessarily altering choices we make. We invariably differ from person to person as we seek to engage in those actions which are most helpful to us as unique human beings.

  For most of us, a lifetime is spent searching for the most satisfying and sensitive insights and actions that can truly tap into ASM'S higher purpose and plan regarding what it means to personally express our sexuality. Inevitably, successes and mistakes occur along the way. We usually express what is most helpful and desirable by visualizing a picture the shape of a heart, accompanied by a sparkle in the eye, a smile on the face, and a tenderness in the touch. However, the strong, pleasurable gravitational pull can include many mistakes and profoundly negative choices.

These are incorporated in such words as: demanding, rough, rape, sexual abuse, prostitution, playing around, fucking, etc. all of which leaves one feeling ultimately unsatisfied and guilt ridden. Inwardly we inherently know which actions are counterproductive to what we truly desire at the deepest level.

Some additional insights I have arrived at over the years would be as follows:

- *Curiosity*. Children are naturally curious about their bodies and genital organs. They wonder where babies come from, they notice anatomical differences between males and females, and many very naturally engage in genital play that includes exhibiting or inspecting the genitals and discovering the pleasurable results of manipulating genital organs. This curiosity and pleasurable discovery is a natural part of ASM and should be simply accepted as a natural part of human growth.
- *Romantic interest*. Sex play, which sometimes includes others, changes as we humans continue to mature. A mysterious attraction begins to develop along with what we describe as a romantic interest. At this stage it is very, very, very important to educate children about the process of procreation including visual diagrams to help them grasp the mystery behind what they have been taught to consider as "their private parts". This needs to happen with both male and female at the level of public education and family relationships so that sex does not remain in the realm of privacy and secrecy about which we are embarrassed or afraid to speak!

- *Masturbation.* Masturbation in moderation is a natural, and safe way to practice self-love and improve health. It produces beneficial results in releasing stress and physical tension. Unfortunately, some cultural and religious taboos have tended to seriously negate its positive and supportive potential. Extensive studies have revealed that approximately 95% of humans engage in some form of masturbation. *We human beings are constructed in such a way that we understand that personal masturbation is pleasurable and relieves tension, but that it also leaves something to be desired.* Sexuality is a part of the total procreative process inevitably involving love and commitment and caring for the other. Singularly expressed it is a positive force, but is never totally satisfying. It's ultimate wonder is realized as the bonding force with the person we cherish the most.

- *Responsible choice.* Because of the great variety of cultural, familial, and uniquely personal circumstances present in each of our own lives, there is no way to create a "one-size-fits-all" approach to the question of what is responsible action for each individual at all times. Rather, the above guidelines as to what responsible sexuality is all about, provides a starting point for assisting in formulating a responsible choice. Each individual must ultimately choose on a personal basis.

Procreation and sexual expression is an automatic, inherent process. It needs to be cherished and understood.

11. Dying to live

In these series of essays I'm asking the question "does nature, explored through natural laws and quantum physics; and Spiritual Awareness, explored through spiritual / religious teachings, have a common, interactive, and crucial message?"

Today we come face-to-face with one of the most significant messages from the natural world; namely, that everything in existence eventually dies. Death is as unmistakable as life itself. The reality of death is first experienced in vegetation as fall and winter appear. It is recognized in the animal kingdom where all animal creatures eventually die, and some even become extinct and disappear from the earthly scene. Museums across the world track cultures and archaeological structures from the past. Everything that is born eventually dies.

But, scientifically speaking, that is not really the end of the story! The truth is that everything is "dying to live". Is it not true that whenever something new is born, something already in existence inevitably needs to die? Think about that! Ponder it! Death is always the precursor to new life. We call it the "food chain". Ultimately, everything is *dying to live*.

- Insects or seeds die to become food for birds.

- Massive groups of sardines die to become food for whales

- A tree dies to become material to create furniture or a home.

Always Something More

- All vegetation dies and becomes compost and nutrient for what is yet to be.

- Dead branches of a tree are removed and even some live branches are pruned in order that the greater good of the whole may be manifested.

- Former tools become antiques and are replaced by more useful, new tools.

True, the *new* energetic life form, animal mineral or vegetable, may be either similar to, or radically different from, what previously existed, but life teaches us that, inevitably, something that was previously unique and wonderful surrenders and dies that "Always Something More" may come into existence. The dying portion inevitably involves decay and suffering. If nature communicates anything to us, whether we like it or not, it is unmistakably apparent that living and dying are part of a recurring process. But, nature also reveals that death is not the end of the story, there is Always Something More. *Everything is dying to live!*

While you can rationally acknowledge that it is so, all forms of animal life, and you as human beings, do everything you can to prolong life and avoid the inevitability of death. This is recognized in the natural world where all forms of animal life are intent on rejecting and fleeing from the reality of "dying to live". In the natural world everything wants to live without suffering, decay and death. Yet, in the final analysis, it is a scientifically verifiable fact that all things die in one form to be reborn in the same or another, modified form.

This inevitably happens amidst the evolutionary process. Cultural traditions, ways of earning a living, orderly patterns of how families live and function; everything in existence is in this eternal, evolutionary transitional process wherein something must necessarily die in order that a greater good and something wonderfully new may come into existence.

Humanly, you pass through an inevitable process of denial, rejection, anger over the loss, bargaining to salvage as much as possible, discouragement and depression; and only after you have gone through some form of this grief process do you reluctantly yield to acceptance and hope. You especially want to eliminate the pain, suffering and agony associated with the dying process. But in the final analysis you surrender and accept the reality that *dying to live* is an inevitable dimension of what life is all about.

Observing nature in its varied forms you discover it to be unmistakably clear that "death" is not the end of the story for there is Always Something More. All things are dying to live

Inevitably, dying to live is a part of ASM. With the passage of time, and the willingness to be changed (continue to grow and evolve into ASM), you eventually learn to accept the reality of death in its ASM perspective and continue the evolutionary process of moving forward into the wonder of that which is yet to be.

Now, it is essential that you *ASMize. S*witch over to the deeper, inherent message contained in varied forms of religious / spiritual awareness through the ages. Hints of the awareness that something must die to live is revealed through sacrificial systems of one form or another from the beginning of time.

Always Something More

- This is the essence of the recorded story of Abraham who was called to die to what is currently satisfying and traditionally accepted, sacrifice the cultural patterns and security of his family, and move with courage into the unknown wherever God might lead; dying to the familiar and courageously moving forth into an unknown tomorrow.
- A hint is found in the fact that most cultures contained a sacrificial dimension of some form. An animal that is pure and unblemished is sacrificed to curry the favor or placate the anger of a superior force known as their God in order that something new and wonderful may occur.
- The story of Exodus in the Old Testament involves the story of the "blood of the covenant" wherein the pain and suffering and loss of the "firstborn" of the Egyptian people occurs in order that God may evolve a special group of people out of their former, limited ways of living and understanding into a new way of life that will enable them to be a blessing to all nations.
- Jesus uses phrases such as these (my paraphrase); "if any man would be my disciple he must deny (sacrifice) himself, take up his cross daily and follow me. Whoever seeks to save his life will lose it and whoever loses his life for my sake and the good news of the kingdom of God will find it!
- Jesus stressed this reality when he told his disciples that he is going to Jerusalem and will be rejected and suffer and eventually die but that he will be reborn again on the third day. Peter exclaims, "God forbid Lord this will never happen to you!" Jesus vehemently

insists that Peter is on the wrong track and needs to radically revise the way he thinks and acts. *His Heavenly Father is insisting that he must lay down his life on Calvary in order that all human kind and all of creation may come to the awareness of self-sacrificing love at the core of all that exists.*

- Meditate on this!! It is at the heart of the Christian faith through the cross on Calvary. In the Christian faith this is expressed in two different images of the cross:

  - The crucifix, which contains the image of Jesus on the cross, symbolizes the necessary and inevitable presence of suffering, rejection and pain in order that self-sacrificing love can redeem the world. In and through Jesus, I myself; ASM, Heavenly Father, was there! His suffering was mine as well!

  - The empty cross clearly emphasizes that there is Always Something More. The cross is empty because of the joy of the resurrection where it is clearly revealed that dying is not the end of the story, but rather that; because of infinite, unconditional, self-sacrificing love, all things are ultimately *dying to live*.

NOTE! ASMize this! *The force of evil and error in the world can only be eliminated when the innocent are willing to die for the guilty. It is only through unconditional love experienced when you offer your life on behalf of others that reconciliation and wholeness can transpire.*

At every stage of the journey Christians refer to as "holy week", which ends in total darkness on the day of the crucifixion, the message is being driven home *you must continually die to the old to most fully live into the new!* It is at the core of what Jesus intended when he established what is known as "the Last Supper". Jesus arranged that, utilizing all the senses: sight, sound, smell, touch, and taste; there would be a meal wherein you'd always remember **that self-sacrificing love is the way of life, and loving and doing good, especially to your enemies,** is the only way to create reconciliation and wholeness.

When you meditate on this overarching reality you discover that Jesus's life and teaching clearly reveals the message that death and resurrection permeates all of life and all aspects of living. You are always *dying to live!*

- You need to Die to materialism and the accumulation of possessions and riches to live into the joy of caring and sharing.

- You must die to your desire to always be comfortable and secure to meaningfully experience the changes that evolution always requires.

- You are called to die to the desire for power and authority in order that you may truly be a servant of all!

- You need to die to the tendency to judge others and condemn them because they're different than you to look deeper and discover some of the infinite diversity that is also a wonderful part of who they really are.

- You must die to vengeance and retaliation to truly care about your enemy and find ways to do good and bless instead.

- You must stop clinging to your life here on earth in order that the Always Something More of what is yet to be can come into existence. Some of you are asking yourself the question, "is there life after death"? If you've really been *ASMizing* you already know the answer: " of course there is"! There is Always Something More!

You are created in My Image and must acknowledge that your task is not to understand the total picture and have all the answers. That's my business! Like Higgs Boson at the heart of each of the trillions of Atoms that constitute your physical body, you are my Enabling Agent and must see yourself, like Jesus, as "Servant of All"; respecting each appropriate part and becoming an evolving, procreative, healing force. If you will, the urge to create and my Wholeness Producing Spirit within you will enable everything to work together harmoniously and for the greatest good.

Do you remember how I said at the beginning that ASM includes paradox and mystery? You human beings have always confronted the reality of death and, by observing nature, developed an image of the hoped-for realities you desire to be present. You incorporate them in the term "heaven" and envision them as being somehow related to the ASM that lies beyond death.

- Countless galaxies and stars permeate the sky and you stand in awe and wonder.

Always Something More

- A rainbow of breathtaking colors penetrates the mist and you are endlessly intrigued.
- Majestic mountains soar upward where sky and earth intersect and cause you to gasp at their majesty.
- Birds migrate unbelievable distances to unknown places at the dictate of changing seasons, and you delight in their freedom.

Everything you have experienced in the natural world around you that delights the eye, thrills the ear, satisfies the touch, entices your taste, and enhances your smell you consider to be "heavenly". It is that wonder-filled Something More than simply will not let you go, but draws you endlessly into the fullness of its embrace.

In like manner, you have always confronted the great unknown beyond death and, observing nature, witnessed realities you desperately hope will not be present. You incorporate them in the term "hell" and envision them as being somehow related to the ASM that lies beyond death.

- Terrifying tornadoes and hurricanes lay waste everything within their path.
- Earthquakes arise from deep within the earth without any consideration for what lies above on the surface.
- Torrential downpours of rain destroy rather than nourish.
- Radiation, in its nuclear form, has the power to destroy all of existence.

Nature is not always kind, beautiful, beneficial, and exhilarating. In equal measure it has the possibility of being painfully destructive. It is this horrible, terrifying form you are referring to by the phrase "hellish". Jesus repeatedly refers to this ultimate reality by the phrase, "outer darkness where men will weep and gnash their teeth." (Matt. 13:40; Matt22:13). As Jesus describes the reality of the final judgment, he includes the fact that all current reality continues to exist in one form or another beyond death.

> Matthew 25 --- The Righteous will evolve into eternal Life and the Unrighteous and evil will evolve into eternal punishment.

That's what it means to be *dying to live*! The "dying to live" potential is not always positive. Exercising our *urge to create* and *necessity of choice* that makes us uniquely human involves inevitable consequences. The choice is always ours, and what we choose lives on into the Always Something More, ASM!

12. ENERGY

Transformative Power = Spirit

The vital force behind all that exists

You are seeking to more fully comprehend the natural world and some of the spiritual insights that are related in *essence* to these elemental realities. This essay considers the *essence* of energy itself.

ENERGY = POWER. Scientists describe the origin of the universe and our planet earth within it by the tern THE BIG BANG. The essence of the "*urge*" that set the creative potential in motion is *Energy*. In our natural world energy is the *power* derived from the utilization of physical or chemical resources. Energy / power is manifested in a variety of ways. Without this '[energizing' reality the material world in which we live would be totally barren and unexciting.

There are two different forms of energy: *Potential* and *Kinetic*

- *Potential Energy* is any type of *stored* energy. It can be chemical, nuclear, gravitational, or mechanical .It's real, but not being tapped or used.
- *Kinetic Energy* is active, working, accomplishing, always in motion.
  - ➢ *Sound* is a form of energy that is associated with vibrations of matter. It is a type of mechanical wave which means it requires an object such as air or water to travel through. Sound originates from the vibrations that result after an object applies a force to another object. It is received by two ears.

Always Something More

- *Light* is a type of radiant energy that we visually perceive with our eyes. In other words, light is power that we can see! Visible light is a small part of a larger reality called the Electromagnetic Spectrum, which contains energy types that travel through space in a wave-like manor.

- *Heat* always refers to the transfer of energy between systems (or bodies), not to energy contained within the systems. In other words, heat is energy, while temperature is a measure of energy.

- *Quantum Physics* is a study of energy fields, such as radio waves, microwaves, infrared rays, ultraviolet rays etc.
  - ✓ **Radio:** Your radio captures radio waves emitted by radio stations, bringing your favorite tunes. Radio waves are also emitted by stars and gases in space.
  - ✓ **Microwave:** Microwave radiation will cook your popcorn in just a few minutes, but is also used by astronomers to learn about the structure of nearby galaxies.
  - ✓ **Infrared:** Night vision goggles pick up the infrared light emitted by our skin and objects with heat. In space, infrared light helps us map the dust between stars.
  - ✓ **Ultraviolet:** Ultraviolet radiation is emitted by the Sun and are the reason skin tans and burns. "Hot" objects in space emit UV radiation as well.
  - ✓ **X-ray:** A dentist uses X-rays to image your teeth, and airport security uses them to see through your bag. Hot gases in the Universe also emit X-rays.

> ✓ **Gamma ray:** Doctors use gamma-ray imaging to see inside your body. The biggest gamma-ray generator of all is the Universe.

RADIATION: Energy / power is then distributed throughout creation by a process we refer to as radiation. Now here is the crucial point to remember regarding the message nature is trying to communicate to us.

> *A law of physics says that kinetic energy can be neither created nor destroyed, but it can change back and forth between forms.*

> *Potential energy* is present, but not actively accomplishing anything! It is present, and stored, but ineffective. *Kinetic Energy* is always in motion and "at work" and needs some form of instruction and direction to become effective.

Now, ASMize. Probe more deeply into the spiritual manifestation of the essence of the energy mentioned above.

The biblical term *ruah* in the Old Testament is the term that is translated as *breath* or *spirit*. Whenever something is occurring that has powerful movement, sparkle, vitality, true and exciting accomplishment we refer to it as being *energized*. Behind that term *energized* is the related term *"spirit"*. When an athlete accomplishes something outstanding we shout, "that's the spirit!" Or when someone is addressing a subject with vitality and enthusiasm we say he is really "getting into the spirit of the thing". Or someone is resisting an evil force in a nonviolent manner we will say, he's acting in the spirit of Gandhi or Martin Luther King or Jesus Christ.

> *It is not the object or goal that is important, it's the energy that inhabits it that is crucial*!

Always Something More

If it's the activated: "*Spirit* of God", the "*Spirit* of Jesus", the "*Wholeness Producing Spirit*", or the Spirit of ASM"; then the ultimate result will be filled with goodness and wonder. It's what Jesus referred to as "The Way". Any other way will not produce what you most deeply desire.

Unfortunately, there are many persons who are religiously involved, substituting certain rules, rituals, regulations, and confessional statements without actively tapping into the *spirit within them.* They refuse to probe more deeply into the ASM of what is occurring. They are not *energized from within.* They lack the enthusiastic involvement that *Spirit* always reveals.

Conversely, those who are seeking the deeper dimensions of Jesus Christ and his life and teachings and strive to daily follow in his footsteps discover that they are *Energized and Empowered* in new and wonderful ways. Life becomes filled with exciting new possibilities. The energy pulsating from God through Christ and his Wholeness Producing Spirit (Holy Spirit) radiates outward to provide an enabling, guiding, and resourcing presence! Paul, One of my disciples, referred to it by saying, "if anyone is in Jesus Christ he is constantly becoming a new creature. The old limitations have passed away and behold, all things are new".

I ask you a significant question! Whatever religious form or tradition you have experienced in the past, is it filled with me, ASM? Always Something More? Are you constantly aware of the Something More that entices you onward and refuses to let you go? Within the Christian faith the season known as Pentecost celebrates the reality that there exists a power-filled, energy flowing from God to enlighten, transform and energize the world. The life and teachings of Jesus releases this energy into its kinetic form.

The power-filled energy of God radiates outward like the heat producing ultraviolet rays from the sun, firing us up with a motivating force that gives us courage and strength to move confidently and with a sense of rightness and peace amidst every force of danger and evil. Jesus followers, who daily apply what he has taught and revealed in his life, discover a transforming energy within that spiritually energizes all of life's experiences and enables them to do things they never dreamed were possible.

This is what Jesus was seeking to emphasize with this phrase, "the kingdom of God is at hand!" You and I are not alone in this world. We are not limited by our admitted inadequacies. Jesus said "I will be with you always, enlightening you, resourcing you, guiding you, empowering you"!

ENERGY is The vital, power-full force behind all that exists. It is the dynamic and active center of who I AM, ASM! You are correct in pursuing energy in every dimension of quantum physics and our natural world. It is also crucial that you pursue this reality of energy as a vital, empowering *Spiritual* force field amidst your daily life. By growing quiet and meditating on your natural world and on the life and teachings of Jesus, *and then actually daring to daily apply what you have been commanded,* you will increasingly discover that you are indeed energized beyond anything you formerly dreamed possible. ASM!

The fact that this energetic forceful urge exists is self-evident. Energy comes in two forms:

> *Potential energy*; which is stored and available, but useless because never tapped and applied;

Always Something More

Kinetic energy; which is always actively at work toward a desired goal!

I am ASM; the *Always Something More* filled with *potential* energy that I desire to shower upon you if only you truly activate it into its *kinetic* expression by desiring and applying.

You alone can must choose to ignore or claim what I offer. The ongoing essays of this journey are being written by you!

ASM

■■

Let's Share
Now I want to enter into a conversation with you.

How can you ask the questions that are on my mind and heart?
Because I am at the core of your very being and participate in and understand everything about you.

Are there more essays still to be written?
Of course, because there is Always Something More. The future essays are being written by you day by day as you enter your ASMize regarding your awareness of Nature and Spiritual realities that are around you.

In what format shall I write these unfolding essays?
- After ASMizing, if you feel led by my spirit you may add to these essays as an appendage when you pass them on.
- If you have been blessed by these essays you can forward them to family, friends and neighbors and tell

them how you have been blessed, encouraging them to participate in the journey as well.
- You can gather a small group around you to jointly share the essence of these essays and then take turns sharing additional aspects of Nature and Spiritual insights that result. This is an especially important option since you were created to be interacting with others and always in a caring and respectful relationship.
- You can decide that, because of your unique nature, these essays don't really resonate with you, and press the delete button to erase.

You are a human being and I have created you to co-create with me on the earthly scene. Personally, I would hope that you have become increasingly aware of my unconditional love that surrounds you and the joy that comes to my heart from continually interacting with you through insights and nudges when you *ASMize*.

In your personal life when you are in love you want to be present and caring for your loved one every day. That is my heartfelt desire as well! I don't just want to interact with you and through you when you are down and out and in the pits, but even more, I want to rejoice and share your joys and daily experiences because I truly do love YOU.

Always remember the first and greatest commandment that is inherent in nature and exists in one form or another it every religious expression. It is known as "The Golden Rule": "do unto others as you would have them do unto you"! And conversely, never do to others what you would not want them to do to."

You cannot love and forgive others properly until you have already learned to accept forgiveness and determined to share it with others and with all of creation. The Always Something More is continuously being written through you.

Hoping you will share with me on the continuing journey. Never forget that you are infinitely and unconditionally loved!   ASM

* * * * * * * * *

I want to extend a word of thanks to my servant, John, for agreeing to be my publishing and distribution agent.

PS… These essays are more easily read, reviewed, pondered, underlined and highlighted in printed book form.   Primarily available from:

Amazon.com/books

www.ingramcontent.com/pod-product-compliance
Lightning Source LLC
Chambersburg PA
CBHW070316230526
45470CB00002B/901